NAVIGATION

SCUBA SCHOOLS INTERNATIONAL

DISCLAIMER:

The information contained in the SSI training materials is intended to give an individual enrolled in a training course a broad perspective of the diving activity. There are many recommendations and suggestions regarding the use of standard and specialized equipment for the activity. Not all of the equipment discussed in the training material can, or will, be used in this activity. The choice of equipment and techniques used in the course is determined by the location of the activity, the environmental conditions and other factors.

A choice of equipment and techniques cannot be made until the dive site is surveyed immediately prior to the dive. Based on the dive site, the decision should be made regarding which equipment and techniques shall be used. The decision is that of the dive leader and the individual enrolled in the training course.

The intent of all SSI training materials is to give individuals as much information as possible in order for individuals to make their own decisions regarding the diving activity, what equipment should be used and what specific techniques may be needed. The ultimate decision on when and how to dive is that of the individual diver.

First Edition
    First Printing, 11/91
    Second Printing, 3/95

Printed in the USA.

SCUBA SCHOOLS INTERNATIONAL
2619 Canton Court • Fort Collins, CO 80525-4498
(970) 482-0883 • Fax (970) 482-6157

CONTENTS

REORDER # 2501NV

ACKNOWLEDGMENTS

Editor in Chief **Laurie Clark Humpal**

Art Director/Illustrator **David M. Pratt**

Photographers **Greg Ochocki**
Blake Miller

Technical Editors **Robert Clark**
Ed Christini
Gary Glark
Chad Carney
Bart Collins
Bob Schaible
Ray Wolf

Proofing Editor **Linda J. Clark**

Graphic Designers **Laurel L. Malenke**
Laurinda M. Baker

FOREWORD

WHY SPECIALTY TRAINING? The answer to "why specialty training?" is quite simple. An entry level course provides you with the motor skills, equipment knowledge, and minimum open water experience needed to be considered a safe diver. However, an Open Water Diver Course does not a specialist make.

Specialty training has two primary objectives: to prepare you for new diving situations, and to improve your level of skill. The SSI Specialty Diver Program provides an excellent introduction to a variety of different diving subjects such as Deep Diving, Boat Diving, Night Diving/Limited Visibility, and many others. These courses are designed to enhance both the enjoyment and safety of each new situation. Specialty training also develops your current level of skill to a much higher level. For example, you may have learned how to follow a compass in your entry level course—SSI's Navigation Specialty course can teach you how to navigate! A specialty course takes you clearly beyond what you may have learned in your Open Water Diver Course.

Each specialty generates its own excitement and opens its own doors. Not every specialty is appropriate to every diver or diving level. Some specialties may not be available in your area, but can be enjoyed when traveling. Every specialty, however, can open new vistas for the diver who wishes to explore all the adventures scuba diving can provide!

The SSI Specialty Diver Program offers the opportunity, through continuing education, to accelerate the learning process that otherwise could only be gained through significant, time-consuming experience. You can quickly prepare yourself to be comfortable in whatever diving situations apply to you personally. Here is a simple, inexpensive way to gain knowledge, experience, safety and recognition, in classes tailored to your specific interests!

THE SSI SPECIALTY DIVER PROGRAM OFFERS THE OPPORTUNITY TO ACCELERATE THE LEARNING PROCESS THAT COULD OTHERWISE BE GAINED ONLY THROUGH TIME-CONSUMING EXPERIENCE.

NATURAL
NAVIGATION

1

CHAPTER 1 :
NATURAL
NAVIGATION

For many years, pilots, seamen and explorers have steered their way through their unique environments using the skills of navigation. Scuba divers can also learn to master their underwater environment by applying many of these same techniques.

When a scuba diver learns to navigate the way to and from a destination, a new sense of confidence and ability is achieved (Figure 1-1). Simply knowing you can find your way back to the boat or shore relieves stress from the dive. The ability to find exciting destinations such as wrecks increases your diving enjoyment. Most exciting, however, is the increased sense of independence; knowing that you can safely dive without the guidance of a dive leader. This independence will open up new diving opportunities and enhance your overall experience!

Figure 1-1   *When a scuba diver learns to navigate, a new sense of confidence and ability is achieved.*

NATURAL NAVIGATION

In this book we will discuss various ways to navigate under water, including some fun ways to increase your ability once you finish this course. The most basic method is the one we will start with first—natural navigation. Most divers are probably already practicing natural navigation to some extent, for it simply involves orienting yourself to your surroundings, and then using these surroundings to indicate direction.

Geographic Formations

As there are mountains and gullies on land, so are there under water. These natural geographic formations can be used for orientation, either as a reference point to start and end your dive at, or as guide to follow along, much as you would follow a river on a hike. Below are some examples of how to use geographic formations while diving.

■ **Walls / Drop-Offs / Ledges:** If you start the dive with the drop-off on your right and swim for 20 minutes, you should be able to turn around and swim with the drop-off (or wall) to your left for approximately 20 minutes and end up where you started.

■ **Coral Heads:** A giant tube sponge, a unique sea fan, or a distinct coral head can be used as a reference point (Figure 1-2). Use a reference point to mark your exit or where you deviated from your course.

■ **Rock Formations:** In areas where rock is the prevalent feature, rock formations can be used in much the same way as coral heads.

■ **Outcroppings:** Both coral and rock reefs may have outcroppings that can be easily spotted. A landmark such as an outcropping (or a coral head) can mark where you descended or turned off course (Figure 1-3). By returning to this reference point you can retrace your course and find your exit point.

■ **Kelp:** A kelp bed can aid in navigation much like a wall, or a coral or a rock formation. The kelp bed's position (relative to you as you enter the water) can aid in determining your position in the water as well as your exit point at the end of the dive.

Figure 1-2 *A giant sponge can be used as a reference point.*

Figure 1-3 *A landmark such as a coral head can mark where you descended or turned off course.*

Using geographic formations to aid in natural navigation will be discussed in further detail later in this chapter.

Landmarks

Man-made landmarks such as wrecks, docks, and buoys can also be used under water to aid in navigation. These landmarks can act either as starting or ending points during your dive (Figure 1-4). Of course, you can also use landmarks in conjunction with the geographic formations already mentioned.

Man-made reference points are also an excellent navigational tool. They could include an old lobster trap or a sunken anchor. These reference points may be a part of an old wreck, or debris that was purposefully sunken. Many times these man-made points will provide better reference than natural formations that can look similar under water.

Figure 1-4   *Man-made landmarks can also be used under water to aid in navigation.*

Illumination

Illumination, or using an artificial light source to guide you, can also be an aid in navigation. This does not only apply to diving at night, because light sources can also be easily distinguished during daylight.

■ **Sun and Light Rays:** Light rays can help you determine the direction of the sun, which will in turn help you determine your direction under water (Figure 1-5). Sun and light rays will be most helpful early or late in the day, when there is a distinct angle or shadow.

Figure 1-5   *Light rays can help determine the direction of the sun.*

■ **Shadows:** A shadow, or darker area, on the the side of a rock or other formation will also help you determine the direction of the sun.

■ **Moon:** When night diving, the moon can often be seen while you are under water. By noting its position while you are still on the surface, you can reorient yourself once you are under water. Do not forget that the moon, like the sun, moves and can change your perspective on direction.

■ **Lights:** Both surface and underwater lights will help you find your way at night. A strong surface light that marks your exit point should be easily seen from under water as you approach it. A strobe or other light source attached to the descent line will help mark your exit point when boat diving. Lights can be seen from some distance away at night. If you would like to find out more about diving at night, SSI offers a specialty course in Night/Limited Visibility Diving.

Bottom Composition

The composition of the bottom will aid a diver in navigation and give clues to direction. At many dive sites, the bottom composition will change as you swim. By noting these changes and when they occurred, you can use this information on your return trip.

- **Slope:** The slope of the bottom may indicate if you are swimming to or away from shore. If you are swimming uphill, or into shallower water, you are probably swimming toward the shore (Figure 1-6).

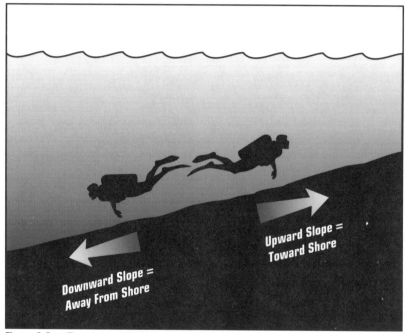

Upward Slope = Toward Shore

Downward Slope = Away From Shore

Figure 1-6   *The slope of the bottom may indicate if you are swimming to or away from shore.*

- **Drop-Off or Wall:** A drop-off or a wall will most likely indicate deeper water away from shore. Be aware that many reefs have inside near shore breaks too.

- **Sand vs. Rock or Coral:** Take note if the bottom composition changes from sand to rock or coral (Figure 1-7). This will indicate that you need to return to the sand to find your exit point.

Figure 1-7   *Take note if the bottom composition changes from sand to rock or coral.*

■ **Sand Ripples:** If you notice distinct ripples in the sandy bottom, try and discern their direction. Because of the wave action, sand ripples will run parallel to the shoreline.

■ **Reef Formations:** Many reefs have a distinct pattern to them. For example, the reef may look like a furrowed field that leads to a drop-off. Note the direction the furrows in the reef run to help track your direction.

The topographical composition of the bottom can give a great many clues to your direction, and provide valuable reference points to use in conjunction with natural and man-made formations.

Water Movement

Just as the waves create ripples in the sand, other movements in the water can help you determine direction.

■ **Surge:** Under water, divers experience surge (the back and forth movement of water perpendicular to shore) caused by the energy of waves. Surge is more noticeable closer to shore or a land mass than in open water.

■ **Currents:** The best way to deal with a current is to swim into it at the beginning of a dive, and then ride the current back at the end of the dive. This will also help you find the direction of your exit point. There are also various localized currents such as longshore, rip, or tidal. If you know these are present at your diving location, it will help you not only navigate, but also dive more safely. Use a current as a directional indicator.

Head into the current, but note when you turn left of the current or back to the right. This is a lot like keeping a recognizable feature in sight on your left or right.

More information on water movement is available from SSI's Waves, Tides, and Currents Specialty course.

Depth

Although divers should constantly monitor their gauges, it is easy to become engrossed in other activities. By paying attention to your surroundings you can notice if you have begun to sink a little deeper than you had planned. Notice that deeper water will be darker and colder, and you may need to equalize your ears again.

Depth can also help you orient yourself under water. By noting your starting depth at your reference point, you will know what depth to return to in order to find that point again. If you are diving along the shore you will also know that the water will get deeper and darker as you swim away from land.

Noise

Since sound travels four times faster under water, even a sound from some distance away can be heard. Boat motors, a compressor on board a boat, an underwater recall system, other divers, and even the crackling sounds heard on the reefs are all noises that can help orient you. A noise will get louder as you approach the source, and diminish as you swim away. By listening closely, you can tell if you are heading in the right direction.

By learning to orient yourself to your underwater surroundings, you can learn to navigate naturally. Your ability to navigate naturally will increase with the visibility, but it will also improve with practice. As you become more comfortable in the water you will gain a heightened awareness of your surroundings, allowing your navigation talents to sharpen.

NAVIGATING BY NATURE

We have discussed some of nature's signposts you can use to navigate without instruments; let us look now at the actual method used to plan and execute a dive with natural navigation.

Orient Yourself on the Surface

One of the most important aspects of natural navigation is planning while you are still on the surface. In order to understand what to expect from the water, bottom structure, and other geographic and man-made landmarks, you will need to ask someone who is familiar with the dive site (Figure 1-8). This will most likely be a dive leader, boat captain, or guide. Because the underwater layout will be similar to that of the surface, features such as coastline, rocks, kelp beds, and wrecks that are visible on the surface will help start the orientation process.

Below is a list of steps to take during surface orientation and dive planning. Remember to ask all questions up front, because it will be too late once you are under water. As with all pre-dive planning, surface orientation will lead to a safer and more enjoyable dive.

Figure 1-8 *One of the most important aspects of natural navigation is planning while you are still on the surface.*

- **Listen in the Pre-Dive Briefing:** Your boat captain or dive leader will give you a description of what to expect from the site.

- **Talk to the Dive Leader if Necessary:** If you didn't receive a pre-dive briefing, or if all your questions were not answered, go ahead and question the dive guide or leader. Or you can ask a diver in your group who is more experienced with the area.

- **Draw a Map or Take Notes:** The easiest way to avoid forgetting information is to write it down. Use a slate to take notes of general directions, or to draw a quick map that you can follow under water (Figure 1-9). Remember to note the direction and distance to geographic formations and landmarks, as well as the current and position of the boat. Remember to note the position of the sun or moon, and the angle of the rays.

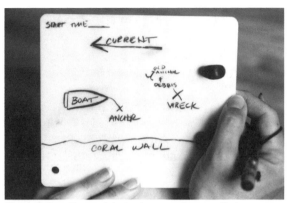

Figure 1-9 *Use a slate to draw a quick map that you can follow under water.*

■ **Make a Dive Plan with your Buddy:**

Once you are both oriented to the dive site, go ahead and plan your dive (Figure 1-10). Decide your general direction, depth, and time in addition to other factors such as minimum p.s.i. (bars) prior to surfacing.

Figure 1-10 *Make a dive plan with your buddy.*

Orient Yourself in the Water

Once you and your buddy are in the water, take some time to orient yourself before you descend. On descent, make sure you are facing the general direction you are planning to swim. This will keep you heading in the right direction once you are on the bottom. Take a moment to reorient yourself before swimming off. If you drew a map, note your position and the bottom features you were told to expect. This extra moment of preparation can make the difference as to how easy, or difficult, it is to find your way back to your exit point.

■ **Study the Surroundings:** After you reach the bottom, take some time to study the surroundings. Now is the time to notice the movement of the water, how the sun or moon appear under water, and the other features such as bottom slope and composition.

■ **Select Reference Points:**

Add any landmarks or reference points you see to your map, or make a mental note. You may even want to mark your reference point by attaching a cylume stick, float, glove, or other object to a small weight that you can leave on the bottom (Figure 1-11). You can also set a shell or other item on top of a rock to make an unnatural marker. Be sure not to harm the reef when marking or creating reference points.

Figure 1-11   *You may want to mark a reference point under water.*

Decide the Direction of your Dive

Once you have oriented yourself to the water it is time to make the final decisions as to your direction, and to note your return course.

■ **Record your Depth:**

By recording your starting depth and the depth of your reference points, you will know what depth to ascend to when it comes time to find your exit point (Figure 1-12). Under water, sea fans or rocks can look alike, this way you will know to look for the giant, purple sea fan at 33 feet (10 metres).

Figure 1-12   *By recording depths it will be easier to find your exit point.*

■ **Note your Direction:** Make a written or mental note of the direction you will travel. If you are swimming along a wall, note if it is on your right or left side.

■ **Note your Return Course:**

As you start to swim toward your destination, occasionally *look back to note how the return trip will look*. Locate landmarks that will help you on the return trip. For example, notice that the wreck is on your left as you swim out, and that it will be on your right on the way back (Figure 1-13).

Figure 1-13 *Notice that if the wreck is on your left as you swim out, then it will be on your right on the way back.*

■ **Note your Time:** Before you take off, make a note of your starting time. One good navigation tactic is to swim along a shoreline or wall for a set time, and then turn around and swim for the same amount of time back. If you don't vary your kick, you should end up in the same general location.

■ **Note your PSI:** Another common tactic used by navigators is the one-third rule of air consumption. The way the rule works is you use one third of your air to swim to your destination, one third for the swim back, and one third for an emergency in case you cannot find the shore or boat.

Now you are ready to begin your dive. Because of your pre-planning, you should be able to enjoy yourself without worrying about finding your way around the dive site. Keep track of where you are, occasionally referring to your map or notes, and make a record of the changes in water movement, depth and bottom composition. This added alertness will also give you a new appreciation for diving. You are now taking the time to notice the subtleties around you: the ripples in the sand, the fish flowing in the surge, and the beautiful geographic formations.

Returning to Your Exit Point

Through your pre-dive orientation and diving alertness, it should be simple to return to your exit point. It may be as simple as reversing your course and swimming for a certain amount of time. You may need to return to a specific depth to search out your reference point. Either way, you should be able to find your way back based on your map or notes. Common pitfalls to avoid are deviating from your planned course without making a note, or taking shortcuts during pre-dive planning.

Reorienting Yourself After Losing Direction

Should you for some reason lose your direction, you will need to stop and try to reorient yourself. The same rule applies to diving as it does to hiking in the mountains: once you realize you are lost, stop before you make yourself even more lost. As a last resort you can always surface and look around, but there are a few things listed below that you should try first.

■ **Stop and Observe the Surroundings:** Once you stop it will be a lot easier for you to detect the subtleties of the surroundings. Try and notice which way the current is flowing. Remember, since you started swimming into the current, you will need to swim with it to get back. Try and see if the sun (or moon) is still in the same relative position as when you started, and if the bottom composition has changed. Basically, you will need to review all of the clues to nature that were discussed in the first section of this manual.

■ **Try and Orient Yourself to Where you Started:** If you started diving near the shore, reef, or kelp, try and remember the last time you saw it. Look up and see if you can see the boat, the anchor line, or any other divers on the surface.

■ **Swim Back to your Starting Depth:** Swim back to the depth you started at, and where your reference point is at. This will give you a better chance of seeing something familiar.

■ Look for Other Divers in your Group:

If you happen to see another buddy team, they will hopefully be able to point you in the right direction (Figure 1-14).

■ Surface and Reorient Yourself:

If you are completely lost or are running low on air, the best choice is to surface (Figure 1-15). Do not forget to make a safety stop at 15 feet (5 metres). If you have enough air, you can drop back below the surface once you are sure of your direction; if not, you can swim on the surface. If you are too tired to make the swim, signal for help.

Figure 1-14 *If you lose your direction, look for other divers in the group.*

With practice and pre-dive planning, you can become quite proficient at natural navigation. Remember, you need to ask all of your questions about what to expect from the dive site *before* you enter the water. Orienting yourself to your surroundings on the surface will give you a better perspective of what to look for under water. Natural navigation will, of course, work better in good visibility, when you know which direction is north, south, east and west, and when there

Figure 1-15 *If you are completely lost, the best choice is to surface.*

are bottom features such as rocks and coral to use as reference points.

When you are diving at night, in limited visibility, or in the open water, you will need to rely on a more sophisticated form of navigation. A compass will provide you with a way to determine and follow precise directions. Compasses and navigation are covered next in Chapter 2.

CHAPTER 1
REVIEW

1. When a scuba diver learns to navigate to and from a destination, a new sense of _____ and _____ is achieved.

2. Landmarks can act either as _____ or _____ points during your dive.

3. The slope of the bottom may indicate if you are swimming to or away _____ _____.

4. Because of the wave action, sand ripples will run _____ to the shoreline.

5. One of the most important aspects of natural navigation is planning while you are still _____ _____ _____.

6. Use a slate to take notes of general directions, or to draw a _____ _____ that you can follow under water.

7. If you are completely lost or are running low on air, the best choice is to _____.

COMPASSES AND NAVIGATION

2

CHAPTER 2:
COMPASSES
AND
NAVIGATION

WHY USE A COMPASS?

The compass is a useful tool that helps the diver maintain a sense of direction when natural navigation is not possible or practical. In these situations, a compass is not only useful, but will add to your overall safety and enjoyment.

A diver does not need to become an expert in navigation to use a compass, in fact, you may have already used a compass at some point in your life. They are frequently used by boaters, pilots, hikers, and anyone who may lose their direction, either in the mountains or open desert. The same applies to a scuba diver in vast, open bodies of water. Let's look at some of the reasons why compasses are so useful to scuba divers.

Direction

The most important reason to use a compass is to maintain direction. In limited visibility, a scuba diver can have a hard time maintaining their sense of what is up or down, let alone which way is north or south. This can also apply to swimming on the surface. Divers may surface in fog or cloud cover, no longer able to sight the boat or shore-line. In this situation a com-pass is critical.

The hardest thing about learning to use a compass is to trust it (Figure 2-1). Much like a pilot must learn to fly by his instruments only, a diver should learn to navigate by his compass. It will be difficult to rely only on your instruments and not your sense of direction. At night or in turbid water, your sense of direction may be completely off.

Figure 2-1   *The hardest thing about learning to use a compass is to trust it.*

Confidence and Safety

Knowing that you can navigate your way to and from any dive site gives a diver an immense sense of self-confidence. More importantly, it adds to your safety.

The ability to navigate allows a diver to plan and execute a dive without having to surface for direction. This is especially important when diving deep or making repetitive dives. Your dive profile will be jeopardized if you continually need to resurface for directions, especially if you don't make a safety stop each time (Figure 2-2). Navigation skills also help a diver to conserve energy and air, while extending bottom time.

Figure 2-2   *Navigation skills allow divers to easily find the ascent line for safety stops.*

Divers can relax and enjoy themselves if they know they can quickly and safely find their way back to their starting point. Your self-esteem will also grow as your buddies realize they can rely on your navigational skills. As your confidence and ability grow, you can begin to explore new waters and increase your enjoyment.

Challenge

As you continually develop your navigational expertise, you will begin to look upon compass work as a challenge. Plan navigation activities to test your abilities and increase your enjoyment, especially when you dive in limited visibility.

Each time you dive, try navigating closer to your destination, while increasing the difficulty of the course. Create a competition of sorts by challenging other divers to a compass run. There are many fun and interesting ways to increase your skill level.

HOW A COMPASS WORKS

Although a compass may look like a complex and confusing instrument at first, it is really very simple. Let's take the mystery out of compasses by examining how and why they work.

What is a Compass and Why Does it Work?

Early compasses were really quite basic. They were no more than a magnetized needle that was kept afloat in a liquid by a piece of straw. As the needle floated, it would point toward the Magnetic North Pole.

While sailors could judge general direction by this crude compass, over the years a more sophisticated system of navigation was developed. The four general directions, north, south, east, and west, were named by sailors for the cardinal points from which the winds blew. Over time, these directions were subdivided until they eventually become a thirty-two point wind dial (Figure 2-3).

As time went on, navigators continued to perfect their ability to navigate by actually designing

Figure 2-3   *18th century round dial.*

a grid system for the entire planet. The earth was first divided horizontally in half to create the Northern and Southern Hemisphere. The equator is the 0 degree starting point. The lines of latitude, parallel to the equator, extend in each direction toward the North and South poles (Figure 2-4).

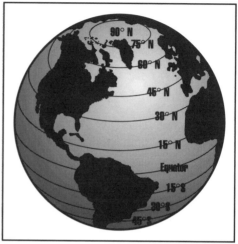

Figure 2-4 *The lines of latitude.*

To divide the planet vertically, an agreement was reached to start the 0 degree line of longitude, or the prime meridian, in Greenwich, England. From this point, the lines of longitude divide the planet equally, 180 degrees in each direction (Figure 2-5). Just as the planet is divided into 2 equal halves of 180 degrees, so is the compass. This is why your compass is set up on a 360 degree system. Actual use of the compass will be discussed later in this chapter.

This grid system of degrees longitude and latitude allows navigators to plot exact courses by use of charts. There are various types of charts available: Sailing charts, General charts, Coastal charts and Harbor charts. Harbor or small craft charts are the only charts that are usable by scuba divers. Your local area may produce diving and fishing charts that can also be used for navigation. By using a chart, a diver can see what to expect from

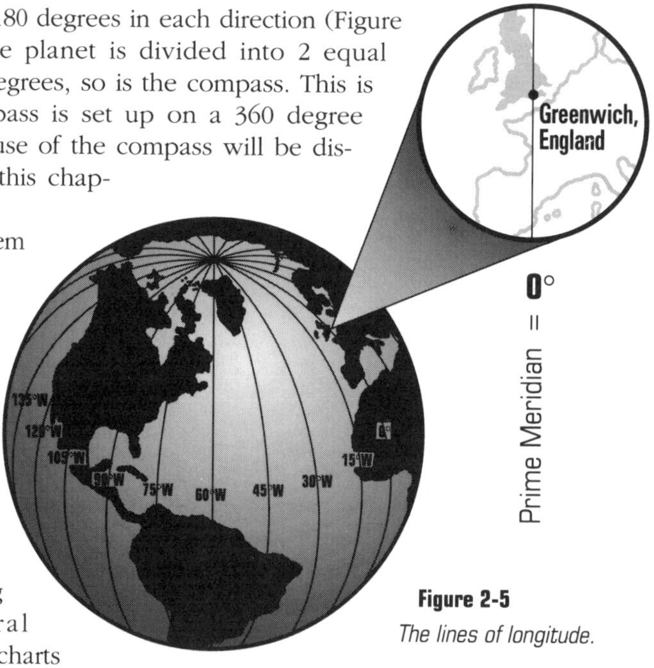

Greenwich, England

Prime Meridian = 0°

Figure 2-5

The lines of longitude.

the dive site, such as depth, hidden reefs, anchorage areas, and landmarks (Figure 2-6). Although charts can certainly be valuable tools for scuba divers, using them for precise navigational purposes is beyond the scope of this course. If you are interested in learning more about charts, consider a boating class in nautical navigation.

Variation and Deviation

As we have already discussed, because the needle of the compass is actually magnetized, it will point towards Magnetic North, however there is a difference between Magnetic North and True North.

Figure 2-6 *Charts can show a diver what to expect from the dive site.*

■ Magnetic North vs. True North:

The Magnetic North Pole actually lies 1000 miles (1609KM) south of the True North Pole (Figure 2-7). For most divers who use a compass, this will never create any problems. However, were you to use a chart (as discussed above) and attempt to follow a specific course, you would actually veer off course due to variation. The longer the distance you travel, the farther you would veer from your

Figure 2-7 *The Magnetic North Pole actually lies 1000 miles (1609KM) south of the True North Pole.*

course. For sailors who travel a great distance variation must be corrected for, but for scuba divers who swim short distances (without the use of a chart), variation is an interesting fact to learn about, but it won't affect your ability to navigate.

■ **Deviation:** Deviation, however, is a problem that all compass users must deal with. Deviation is created when a metal object comes close to or in contact with the compass. This source of metal can actually attract the magnet instead of letting it point north. As you can easily guess, this will quickly lead a diver off course. The easiest way to combat deviation is to hold the compass away from any metal objects such as a tank, knife, or possibly a regulator. Before diving, check to see if any of your equipment is causing deviation. If so, hold your compass at arms length to try and find the best position to prevent any deviation.

Deviation can also occur from swimming over large mineral deposits such as metal ores, or when swimming near wrecks. Stay as far away from these metal sources as you can when you are using a compass.

SELECTING A COMPASS

Now that we have discussed how a compass works and the factors that affect it, let us look at the parts of a compass and how to select a quality one.

Parts of a Compass

A compass is actually a simple device with few moving parts. Although compasses come in a variety of shapes and sizes, they all consist of the same basic components (Figure 2-8).

■ **Magnetic Needle:**

In order for a compass to work it must have a magnetic needle to point north.

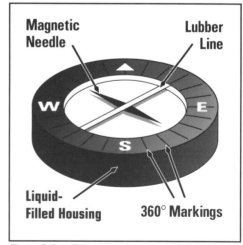

Figure 2-8   *The parts of a compass.*

■ **360 Degree Markings:** On some compasses these markings may come on a compass card that floats in liquid; on others they may come on the outside bezel, or fixed on the case of the compass.

■ **Liquid-Filled Housing:** The compass should be liquid-filled so the needle can float freely.

■ **Lubber Line:** The lubber line is a solid line or an arrow permanently attached to the face of the compass housing. It designates the direction you should swim to stay on course.

Selecting a Quality Compass

Although compasses have the same basic parts, the quality of the compass will vary. Of course, the higher the quality of the compass, the longer it will last, the easier it will be to use, and the more accurately it will read. Below are some of the features to look for in a quality compass (Figure 2-9).

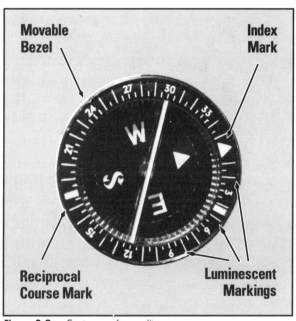

Movable Bezel

Index Mark

Reciprocal Course Mark

Luminescent Markings

Figure 2-9 *Features of a quality compass.*

■ **Magnetic Needle:** The needle should be able to move freely, without coming in contact with the compass housing, even if the compass is tilted.

■ **Movable Bezel:** A quality compass will have a bezel that can be rotated. Depending on the style of compass, the bezel may rotate the degree markings or the index mark. The bezel must be easy to rotate under water, yet stay in place once it is set.

■ **Index Mark:** The mark should be on the face or bezel of the compass, and it should be both bold and luminescent so it can be easily seen. The index mark may be a line, an arrow, or a set of bars. To maintain a course, the north arrow will be lined up with the index mark.

■ **Reciprocal Course Mark:** This mark will be directly opposite the index mark on the compass. This allows you to set your reciprocal course by simply turning your body 180 degrees until the north arrow is lined up with this mark. You do not have to reset the bezel or figure your new setting.

　If your compass does not have a reciprocal mark, you can make one by engraving a mark on the bezel (exactly 180 degrees opposite the index mark) with a sharp pointed object, and then filling in the mark with a little white, waterproof paint.

■ **Large, Easy-to-Read Markings:** The degree marks should be bold and easy to read when held at arms length.

■ **Luminous Dial:** The markings should be luminescent so they can still be read in low light.

■ **Oil Filled Housing:** The compass should be filled with a liquid, such as oil, that will not freeze. Oil will also slow the needle's movement so it will remain steadier.

■ **Resistance to Corrosion:** Any metal parts will eventually corrode from salt water. A quality compass will not have any exterior metal parts, but will be made of plastic.

　Your local, professional, SSI dive store should be able to help select a quality compass that will last for many years.

TYPES OF COMPASSES

There are various types of compasses available. The type you select will depend on how much you are willing to spend, and how you intend to use it. As stated earlier, a higher quality compass may be more expensive, but it will be more accurate and easier to use.

The three general categories of compasses are the top-reading, the side-reading, and the watch band.

Top-Reading Compass

Top-reading compasses are the most popular type, because they come in such a wide variety of styles and price ranges (Figure 2-10a). A good top-reading compass should have certain features, the most important being a movable bezel so you can set your starting course, and then reverse it for your reciprocal course, and an index mark, or set of bars, that is easy to keep centered on the north arrow. The compass should also be fairly sensitive so it can quickly pick up any deviation from course.

A top-reading compass can also come in two different styles, direct reading or indirect reading, depending on where the degree markings are located. A direct reading compass has the numerical degree markings right on the

Figure 2-10a   *Top-reading compass.*

bezel, so they rotate when the bezel is moved. An indirect reading compass has the numerical degree markings right on the compass housing, and the index mark instead rotates when the bezel is moved. An indirect compass is the most desirable model and is the easiest to use, so, for the purposes of our manual, we will use indirect compasses in our examples when we discuss how to use and set a compass.

Less expensive top-reading compasses will not have all the quality features, some may not have a lubber line to follow, and many will be very slow reacting.

The benefits of a top-reading compass are that it is usually easier to keep on course. With a top-reading compass, all you must do is keep the north arrow and the index bars lined up, and you can reach your destination.

Side-Reading Compasses

The best compass you can buy is one that is both top-reading and side-reading (Figure 2-10b). A side-reading compass differs in that you do not have to look down at the top of it to set it and stay on your course. To set this compass, you must line the degrees up in the window on the side of the compass, instead of lining up the north arrow and the index mark. Many people prefer using a side-reading compass because they find it easier than learning to properly set the bezel and index mark of a top-reading compass.

The benefit to this system is that it is easier to run a compass course in which you must follow specific degrees, such as when you know the wreck lies at 240 degrees.

Figure 2-10b *Side-reading compass.*

Side-reading models also allow a diver to accurately sight over the compass so you can navigate, while still seeing where you are going. This compass will most likely be used when swimming on the surface and sighting the boat or shore.

The difficulty with this system is it is easier to get "off course" if you only adjust your course by moving your arms, instead of realigning your entire body. Getting an accurate reading from a compass will be discussed later in this chapter.

Both top-reading and side-reading compasses are available in a console or as a wrist mount. A wrist mount compass straps onto your wrist like a watch. With a console unit the compass is mounted as part of your console with your other instruments. The benefit to a console is that you will never forget your compass, and it is easier to sight over a console for a more accurate reading. A wrist compass is popular with divers who may already own a console to which a compass cannot be added, or for divers who only use a compass on occasion. Some manufacturers also make a compass that can be attached to the high pressure hose of your instrument console.

Watchband Compasses

The least popular style of compass is the watchband model (Figure 2-10c). Because it simply hooks onto your watchband, it is compact and inexpensive. However, it is also somewhat inaccurate and should only be used for general direction control, such as when you know the shore lies to the west. This compass should not be used at night or in limited visibility, or anytime accurate navigation is critical.

Figure 2-10c   *Watchband Compass.*

CARING FOR YOUR COMPASS

Compasses require little care and maintenance because they have almost no moving parts. You should wash your compass in fresh water after every dive, just like you do with all your other equipment (Figure 2-11). Make sure it is free of all sand, salt and mud. You may want to even soak it overnight occasionally to make sure all the salt is dissolved.

If your compass can be disassembled, you may want to take it apart occasionally to soak it in fresh water and clean out any hidden dirt and salt. Lubricate the mechanisms where possible and wipe off any excess.

Figure 2-11   *You should wash your compass in fresh water after every dive.*

Because compasses are fragile, care for them as you would your other instruments. Pack them in a protective bag or inside a neoprene bootie for both traveling and storage. Remember never to store your compass in direct sunlight. The sun (and heat) may cause the inner capsule to expand, resulting in permanent damage. Always store your compass after it has been cleaned and lubricated so it will be in good working order for your next dive trip.

USING A COMPASS

You may have already used a compass at some point in your life, either while hiking or in a boating or flying course. However, even if you have no experience at all, using a compass is really very simple. For people who would like to go on and become advanced navigators, compass use can become as complex as they would like to make it.

Before attempting to use a compass under water, you may want to try it several times on land. This practice will make it much easier to see if you are doing it right, before you go under water and have less visibility, and more interferences.

For our purposes let's start at the beginning with the basics of a simple reciprocal course, a course that goes from point A to point B, and back to point A (Figure 2-12).

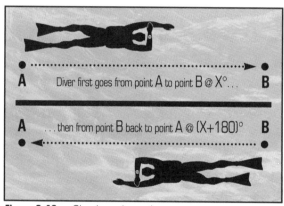

Diver first goes from point A to point B @ X°...

...then from point B back to point A @ (X+180)°

Figure 2-12 *Simple reciprocal course.*

Establishing A Reference Point

The first step is to select the reference point you will be diving to. It may be a wreck that your boat captain has told you about, or it may be a bay on the other side of the lake. Either way, you will need to sight your reference point before you begin. As we mentioned earlier, the more information you have about the site while still on the surface, the easier it will be to navigate naturally under water. The same thing is basically true when using a compass. If your dive leader points you in the direction of the wreck, all you need to do is to line yourself up in the same direction and take a compass heading. Now, when you are ready to go diving, make sure you begin in the same place you took your original heading or you will be off course.

Take the Compass Heading

To take a heading, simply point the lubber line towards your reference point (your direction of travel), and wait until the north arrow stops. Now all you must do is line up the index marks on the north needle (Figure 2-13). After you are in the water and ready to descend, just realign the north arrow with the index marks and you will know you are heading in the right direction. As long as you swim in line with the lubber line, while the north arrow and index marks are lined up, you will be heading in the right direction. If you are using a side-reading compass, you will need to keep the degree marks lined up.

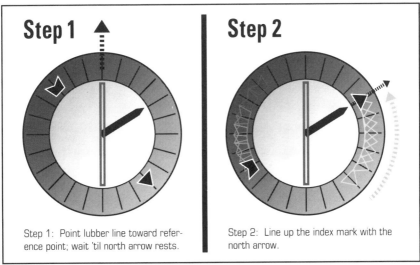

Step 1: Point lubber line toward reference point; wait 'til north arrow rests.

Step 2: Line up the index mark with the north arrow.

Figure 2-13   *Taking a Compass Heading.*

If you are boat diving, remember to let the boat swing into its final position before setting your compass. You may want to reconfirm the direction of your destination before entering the water. This way you will always know you have an accurate compass setting. For more accuracy in a current, you may want to set your compass once you reach the anchor line. Sometimes the boat will swing off of its course due to the wind blowing from another direction.

If you are beach diving, realize that you must enter from the same point where you took your compass heading. If you enter 20 feet (6 metres) up the beach, you will still be swimming at the same angle, but will arrive 20 feet away from your destination (Figure 2-14).

It is a good idea to make a written note on your slate of your compass setting (the degrees) so you can find your way back to the boat even if you accidentally move the bezel and reset your compass. If you know the compass heading, you will always be able to reset the compass and continue your course. It will also make it easier to calculate your return course if you have a slate to write on.

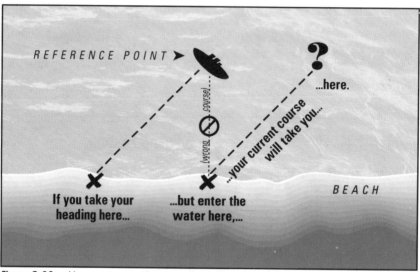

Figure 2-14   *You must enter the water from the same place you took your compass heading or you will be off course.*

Following a Compass Course

No matter how accurate your compass setting, you will never reach your destination if you do not swim where the compass points. There are several factors to consider when using a compass: keeping the compass level, arm position, body position and depth.

■ **Keep the Compass on a Level Plane:** As you hold the compass it must be level both from front-to-back, and from side-to-side. If the compass is tilted too far to one side, the compass card or needle may drag on the housing and prevent an accurate reading.

■ **Arm Position:** Arm position is a big part of getting an accurate reading; your position will vary depending on the type of compass you are using. If you are using a wrist compass you can take the hand with the compass on it and place it on your opposite forearm (Figure 2-15). This position can become tiring on your arms on a long compass run.

A better position may be to take the compass off your wrist and hold it in both hands (Figure 2-16). These positions will also work if you have a console mount compass. If your compass is side-reading, you will need to extend your arms straight out in front of you so you can sight over it. If the compass is top-reading you can lock your elbows into your sides so you can look straight down on it (Figure 2-17).

As mentioned earlier, a top-reading compass is easier to read accurately, and that is because of this difference in arm positions. By locking your arms at your sides, you will have to move your entire body to get back on course. If you are simply holding the compass out in your hands, as with a side-reading compass, you can easily adjust only your arms to put the compass back on course; your body however, is still swimming off course.

Figure 2-15 *Arm position with a wrist compass.*

Figure 2-16 *You may want to take the compass off your wrist and hold it in both hands.*

Figure 2-17 *If the compass is top-reading you can lock your elbows into your sides so you can look straight down on it.*

■ **Body Position:** This leads us to the next factor of body position. As you must keep the compass on an even plane, so must you keep your body on the same plane. In fact, improper body position is the biggest reason divers can't make an accurate compass run. You need to think of the lubber line as extending straight through your body. Your back must be straight, and your feet directly in line with your head (Figure 2-18). If your arms are pointing one direction and you are kicking in another, you will not stay on course.

The best way to stay in the proper position is to designate one member of the buddy team to be the lead diver. If you are navigating, have your buddy swim slightly behind and to one side of you, keeping an eye out so you do not swim into any obstructions, and to make sure you maintain perfect body position. This will also help minimize the magnetic effect of your buddy's equipment to prevent deviation.

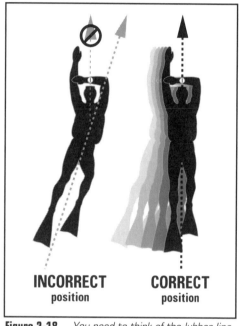

INCORRECT position

CORRECT position

Figure 2-18 *You need to think of the lubber line as extending through your body.*

■ **Constant Depth:**

In order to reach your specific destination, you may need to maintain a constant depth. Have your buddy monitor your depth and check to see if you are swimming up or down (Figure 2-19). You will need

Figure 2-19 *One diver should monitor depth while the other navigates.*

to keep your body flat and swim on an even plane. This will be difficult if you are swimming along the bottom, for you will tend to keep an even distance from the bottom, versus swimming at a constant depth. If your compass is in a console, it will also be easier for you to monitor your own depth gauge (and air pressure) at the same time.

■ **Sight a Landmark:** When swimming a compass course under water, sight along your intended course at the outer limits of the visibility and pick out two prominent objects that are in line with each other as well as your course. Swim to these objects, keeping them aligned. This is not only easier than "following the needle," but allows you to do a bit of sight-seeing along the way. Just before you reach the nearer of these marks, pick out two more ahead of you. Repeat this as often as necessary.

■ **Board Mount the Compass:** If you are interested in competitive navigation, or need precise readings, you may want a board-mounted compass. Some are simply slates with holes cut to fit a wrist compass. Others are designed to hold both an expensive, high quality compass along with a depth gauge. By mounting the compass on a board it is easier to hold steady, and the lubber line can be extended onto the slate for more precise sighting. The longer the lubber line, the easier it is to detect subtle changes in your course.

As you become more comfortable with using a compass, it will be easier for you to maintain the proper body and arm position. With a little practice it will feel more natural, take less thought, and become second nature.

After all this work you should reach your destination right on target. The best thing is, you can relax and enjoy the site because you don't have to worry about the return trip. You know just how long and how much air it took it took you to swim here.

Plotting a Reciprocal (Return) Course

When it is time to return to your exit point (or go from point B back to point A) you will need to plot your reciprocal course. The easiest way to do this is to use your reciprocal course mark on your compass. This mark is exactly 180 degrees opposite the index mark. To set your new course, all you need to do is turn your body until the north arrow is lined up with the reciprocal mark instead of the index mark. You have now reversed your course 180 degrees without moving the compass bezel, or doing any calculations. You can also look at the opposite end of the lubber line and read the degree marking there, then reset that heading to the forward end of the lubber line.

If your compass does not have a reciprocal mark, you will need to calculate and reset your new course. If you have written your initial compass heading on your slate, it will be easy to calculate your return setting.

To turn around you will need to change your course by 180 degrees. This means you will need to either add or subtract 180 degrees from your starting setting, not to exceed 360 degrees (Figure 2-20). One simple rule to remember is that if your original heading was less than 180 degrees: add. If your original heading was more than 180 degrees: subtract. Once you calculate the new heading, first turn the bezel to that spot, and then turn your body to realign the north arrow with the index mark. Now you should be able to return to your exit point right on course.

Step 1: Diver aligns index mark with needle to set course, and notes current heading of 70°.

Step 2: Diver rotates bezel 180° to (70 + 180) = 250°, aligning reciprocal mark with needle.

Step 3: Diver turns around, facing new heading of 250°, 'til needle and index mark realign.

Figure 2-20   *3 steps to setting a compass for a reciprocal course.*

Before we continue on to Chapter 3, let us quickly review the 10 steps for plotting a simple reciprocal (or 2 way) compass course.

1 *Establish your reference point.*
2 *Take the compass heading.*
3 *Make a note of the heading on a slate.*
4 *Realign the north arrow after entry.*
5 *Swim with proper body and arm position.*
6 *Take time to enjoy the dive site.*
7 *Calculate your return setting by adding*
or subtracting 180 degrees.
8 *Reset your compass for your reciprocal course.*
9 *Turn your body to realign the north arrow.*
10 *Swim with proper body and arm position.*

Now let us move on to Chapter 3 and look at how to combine natural navigation with compasses to further increase your flexibility and diving enjoyment.

CHAPTER 2
REVIEW

1. The most important reason to use a compass is to maintain _____.

2. The _____ _____ _____ actually lies 1000 miles (1609 KM) south of the True North Pole.

3. _____ is created when a metal object comes close to, or in contact with the compass.

4. The index mark may be a line, arrow, or a set of bars. To maintain a course, the _____ arrow will be lined up with the _____ _____.

5. The best compass you can buy is one that is both top-reading and _____ – _____.

6. As you hold the compass it must be level both from front-to-back, and from _____ – ___ – _____.

7. Improper _____ _____ is the biggest reason divers cannot make an accurate compass run.

COMBINING
NATURAL & COMPASS
NAVIGATION
3

CHAPTER 3:
COMBINING NATURAL & COMPASS NAVIGATION

On many of your dives, especially in clear water or at familiar dive spots, you will be able to apply your knowledge of natural navigation to pilot to and from your dive site. Even on these perfect days, however, it is reassuring to know that you have both a compass and the skill to use it, just in case.

By learning to combine the skills of natural and compass navigation you can increase your skill and confidence, while increasing your flexibility. You can dive by your natural senses when you are able to, and dive by your compass when you need to.

By learning to combine many skills and techniques you will be more qualified to dive in the "real world," for in reality, you will never know what to expect from each diving day and each dive site. Your navigation abilities will allow you to relax and have fun, because you can easily find your way, without spending the entire dive looking at the compass. So let us take a look at some of the times you may want to combine natural and compass navigation.

TRAVELING TO AND FROM YOUR ENTRY POINT

Even if you are planning to use natural navigation, you may want to take a compass heading for your destination before entering the water, just as a back up (Figure 3-1). If you get in the habit of doing this, it may come in handy. Even a perfect day of diving can have surprises. The visibility can worsen due to storms and other natural causes, both under water and on the surface. The water could get cloudy and dark from rain, or you may ascend into a fog bank or cloud cover and not be able to see the boat or shore.

Figure 3-1   *You may want to take a compass heading for your destination before entering the water, just as a back up.*

If you are diving off a beach, in an unfamiliar area, or in low visibility, you may need to take a compass heading so you can find your way to and from your destination. However, if you are diving a specific area such as a wreck or coral head, you may be able to use natural navigation once you reach the destination (Figure 3-2). This meth-

Figure 3-2   *You may need to take a compass heading so you can find your way to and from your destination.*

od allows you the freedom to explore the site, but provides the confidence of direction control to and from your entry point.

A SIDE TRIP FROM THE DIVE SITE

A different situation may be when you use natural navigation to and from the entry point, such as when you are diving on a wall, or along a shoreline. As long as you have your reference point you will have a sense of direction, but should you decide to swim into the open water or away from your reference point, you should take a compass heading (Figure 3-3). This will allow you to return to your reference point so you can find your way back to the shore or boat.

Figure 3-3   *Should you decide to swim into the open water or away from your reference point, you should take a compass heading.*

GENERAL ORIENTA- TION

One simple use of a compass is for general orientation, or when you just want to know the general direction of something (Figure 3-4). You may not set your compass before diving, but you may notice that the shore is to the west, the boat is to the north, and the dive site is to the east. Now, if you get confused under water you can look at your compass and know that if you head north, you will be swimming in the right direction to find the boat.

Figure 3-4   *One simple use of a compass is when you just want to know the general direction of something.*

In these situations, the compass is more of a reference tool than a navigational aid. You can also use the compass to test your sense of direction. Decide which way you think is west and then check your compass to see if you are right. Through these simple tests, you can begin to sharpen your natural sense of direction, and orient yourself to different dive sites for future reference.

FOR SURFACE ORIENTATION

There may be times when you are naturally navigating that you completely lose your direction. In these instances, the best thing to do is surface and take a look around (remembering to make your safety stop at 15 feet / 5 metres first). If you are low on air, you will need to swim to your exit point on the surface. Unfortunately, this can be hard work and even dangerous if there are heavy waves, currents, or boats in the vicinity.

If your air supply allows it, the best thing to do is to take a compass heading and drop below the surface for the swim back. This way you can drop down far enough to be safely out of the way of boats and waves. You will conserve energy and have a more pleasurable swim back

RETURNING TO A DIVE SITE

Often we find a really great dive site only to discover that when we attempt to relocate it, we cannot! By combining compass and natural navigation you will be able to return to within a few feet of any site you get a fix on whether you swim to it or are using your boat. The key is quick triangulation.

When you find a site you will want to return to, surface directly, swimming into the current to maintain your position immediately above the spot. Once you are on the surface, but before you drift away, take a compass bearing on each of two prominent landmarks ashore. Ideally these landmarks should be 90 degrees apart (Figure 3-5). Sketch these marks on your slate with both the bearing and the reciprocal bearing so you can later transfer them to a more permanent record in your log book. Lines drawn along the compass bearings from these two marks will intersect at your new dive site.

Another method frequently used to relocate a dive site is "Topside Natural Navigation." With this method, instead of just compass bearings, which you will use to get close to the site, you use two additional sets of marks, one behind the other (front and back), called a "range." Using two ranges at least 90 degrees apart will allow you to get a "fix." One of these ranges will be your north/south range, the other your east/west range. A line drawn through the marks on these ranges will intersect right where you want to be (Figure 3-6).

Now, when you want to return to this site in the future, you will simply need to realign the two ranges. Start by realigning the front and rear marks on one side, for example, your right side. Once you have a "fix" on this range, you can determine the direction you must move to align the set of marks on your left. There is only one point where both marks in both ranges will be aligned and that is your dive site. If you use a "sharp" marks such as a radio tower, the edge of a tall building, or a church steeple that are as far apart as practical to make a "range," and if the two ranges are 90 degrees or more apart, you can return to a spot literally within inches. Remember

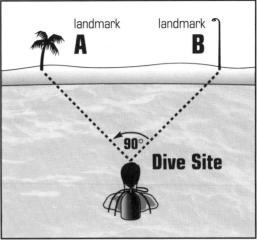

Figure 3-5 *When you find a site you want to return to, take a compass bearing on each of two prominent landmarks ashore.*

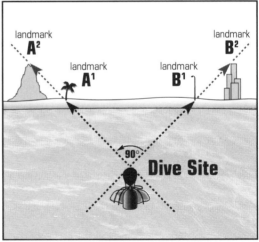

Figure 3-6 *Using two ranges at least 90 degrees apart will allow you to get a "fix." There is only one point where both marks in both ranges will be aligned—that's where your dive site is.*

to start your descent a little "up-current" to allow for drift during descent.

Even when you are able to navigate by nature, a compass provides a sense of security and confidence; the confidence that you can avoid getting lost. By taking a compass heading each time before you enter the water, and each time you stray from your known path, you will always be able to find your way back to where you started. Compasses provide a good general orientation tool and should be used in combination with natural navigation anytime you want that extra sense of confidence on your dive.

Now that we have discussed the difference methods of orientation under water, we will go one step further in Chapter 4 and discuss some special situations that require special navigation techniques.

CHAPTER 3
REVIEW

1. Even if you are planning to use natural navigation, you may want to take a _____ _____ for your destination before entering the water, just as a back up.

2. One simple use of a compass is for _____ _____, or when you just want to know the general direction of something.

3. There may be times when you are naturally navigating that you completely lose your direction. If your air supply allows it, the best thing is to take a _____ _____ and drop below the surface for the _____ _____.

4. _____ provide a good general orientation tool and should be used in combination with _____ _____ anytime you want that extra sense of confidence on your dive.

SPECIAL NAVIGATION SITUATIONS

4

CHAPTER 4:
SPECIAL NAVIGATION SITUATIONS

When conditions are good, your ability to navigate accurately, whether by compass or nature, increases. But what happens when you must deal with currents and poor visibility? How does navigating from a boat differ from navigating from shore? In this chapter we will look at these special situations and the different techniques you can use to improve your navigational abilities.

LIMITED VISIBILITY

Diving in limited visibility can be an adventurous experience on its own, but attempting to navigate through the turbidity adds an additional challenge to the diver (Figure 4-1).

The difficulty with navigating in limited visibility is that you can never see the "whole picture." Your vision is confined to the small area that you can see in, whether it is 3 inches or 30 feet, because limited visibility does not necessarily mean you can't see anything, it just means the visibility is less than it normally would be, or less than optimum.

Figure 4-1   *Attempting to navigate through turbidity adds an additional challenge to the diver.*

Because it is difficult to see much of the area, let alone any reference points or landmarks, it makes it very difficult to use natural navigation. Only experienced divers who are very familiar with a dive site could get by with natural navigation. This leaves compass navigation as the alternative.

Instruments Prevent Disorientation

When you are diving in turbid water, your instruments may provide your only sense of depth and direction. Because you cannot see the surface, bottom, or anything except what is directly in front of you, it is difficult to use your natural sense of direction.

In very turbid water, you may not even be able to tell which way is up or down, let alone which way is north or south. If you begin to experience sensory

Figure 4-2   *By focusing on your instruments you can give yourself a sense of direction, and always know which way is up and down.*

deprivation, or you get confused about direction, you may begin to get disoriented, and even experience vertigo. However, by focusing on your instruments you can give yourself a sense of direction, and always know which way is up and down (Figure 4-2).

If you would like to learn more about diving in limited visibility, SSI offers a specialty course in Night/Limited Visibility Diving. Check with your local SSI store for course information.

Compasses Increase Safety, Confidence and Enjoyment

Diving in limited visibility can be stressful and even somewhat dangerous to an unprepared and untrained diver. A compass will help lower your stress by providing direction control. This added confidence will also make the dive safer and more enjoyable. As has been stated many times, if you have the confidence to navigate, especially in limited visibility, you can relax and enjoy the dive.

NIGHT DIVING

Night diving is very similar to limited visibility in that it is difficult to see the whole picture. Of course, if you are night diving in clear water your vision will be better than in turbid water, but at night your vision is limited by the distance your light will travel and by the natural illumination of the moon.

Navigate by Lights

A compass is particularly handy at night because you must be more accurate about reaching your exit point (Figure 4-3). You may be in the general vicinity and just not be able to see it. This is why lights are so important to night diving. A light will be seen from some distance, even in turbid water,

Figure 4-3 *A compass is particularly handy at night because you must be more accurate about reaching your exit point.*

and will help mark important objects such as ascent lines and exit points. Once you are on the surface, you can simply take a compass heading for the shore light at your exit point and drop back below for the swim back.

All divers should mark themselves with both lights and cylume sticks, but they should also mark their exit point, the ascent line to the boat, the dock, or any other important objects that could be used as landmarks, or that should be pointed out as obstructions (Figure 4-4).

Figure 4-4 *At night divers should mark themselves with both lights and cylume sticks.*

Lights are also important for reading your instruments at night. Some compasses may glow in the dark, or may have a luminescent face that will glow for a period of time after a light has been shined on it. This is a good feature if you are planning on navigating at night.

Night diving can be an exciting challenge, but good navigation skills will help make it a successful experience. Navigating can be tricky because you will need to read your compass and hold your light, yet watch for obstructions, all while you are trying to admire your nighttime surroundings. Your buddy will be able to help take some of the burden off of you, and it will also help to dive sites that you have also dived during the day. This way you will be at least somewhat familiar with the area. If you are interested in more information on night diving, contact your local SSI retailer.

DEEP DIVING

When deep diving, the last thing a diver should do is get lost, or surface for directions. The diver must be able to navigate back to the ascent line without undue delay, because time is of the essence when deep diving (Figure 4-5). Your original dive plan of depth and bottom time must be adhered to or you may find yourself in a decompression situation. If decompression stops were not figured into the dive plan, air could run low.

Deep divers should have additional tanks tied to the ascent line in case extra air is required for the 15 foot (5 metre) safety stop or emergency decompression. This is why it is so important that the diver is able to find this line while still under water. It would be unsafe and impractical to have to surface to look for the boat without making your safety stop.

Figure 4-5   *When deep diving, the last thing a diver should do is surface for directions. The diver must be able to navigate back to the ascent line without undue delay.*

All these factors lead to the conclusion that accurate navigation is important to safe deep diving practices. To learn more about deep diving, contact your local SSI store about a Deep Diving Specialty.

BEACH DIVING

It is, of course, always a good idea to dive in as calm of waters as possible, but sometimes either tidal or localized currents are present. The more advanced a diver becomes, the more likely he or she will encounter currents.

Currents cause water to flow towards and away from shore, as well as parallel to the shore, and can cause resistance for a diver swimming into and across the current. Currents are a problem for boat divers and beach divers because they cause divers to drift off course, making it difficult to navigate to and from the exit point.

When navigating from a beach, a water condition that may cause some problems is a localized current, or a current that runs near the shoreline. Longshore currents flow roughly parallel to shore and are generated by waves and/or wind which approach the shore at an angle. So, for example, if you are diving on the west coast of the United States, the longshore current will flow to the north or south.

Because longshore currents will flow in one general direction, you also will drift with the current while diving. In order to return to a specific point, taking into account the longshore current, you will need to adjust your course. Begin by taking a compass bearing on your destination; let's say that it is perpendicular to the shore at 270 degrees. Now, if the current is flowing north you will actually need to swim at a southerly angle (260, 240 degrees, etc. depending on the strength of the current) to compensate for the northward drift of the current (Figure 4-6a). When you return you will still be drifting northward, so instead of the direct reciprocal of 90 degrees, you would use a course of 100 to 120 degrees, etc. to counteract the current (Figure 4-6b).

To simplify, think of the current as flowing from your left to your

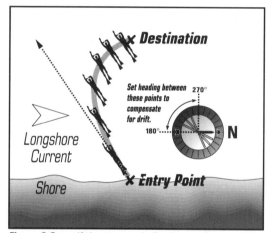

Figure 4-6a *If the current is flowing north, you will actually need to swim at a southerly angle to compensate for the northward drift of the current.*

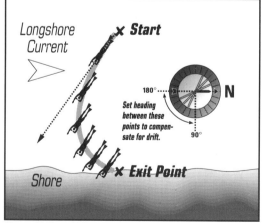

Figure 4-6b *When you return you will again need to swim at a southerly angle to counteract the northward drift of the current.*

right. You would have to swim slightly left of your compass bearing on the way out, and slightly right of your compass bearing on the way back.

Any time waves reach the shore over a shallow reef or a sand bar, that water usually returns to the sea via a rip current (Figure 4-7). Depending on the size and frequency of the waves approaching shore, the direction from which they approach, and the shape of the shoreline, rip currents of varying direction and strength result.

When shore diving, you can enter the rip current and allow it to float you out beyond the breakers. When you return, however, you must choose an exit point away from the rip current, since it is moving outward toward sea. This cre-

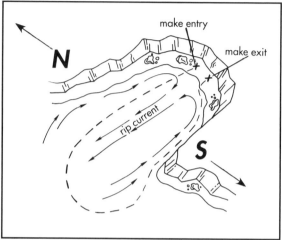

Figure 4-7   *Rip current forming in a cove.*

ates an opportunity to use your navigational skills since your entry and exit points are different. When you enter, take a bearing on the location of the rip so you can avoid the area on your return.

Another method is to avoid the rip current altogether. After locating the rip, look up and down the beach for the best place to enter and exit. Use those areas when plotting your compass bearing for your dive.

The methods discussed in this section are general tips to help beach divers navigate. To accurately navigate in a current, especially when boat diving in the open water, is an advanced skill that takes years of practice. The next section on currents is optional depending upon the experience of your instructor and whether this activity is applicable to your diving region.

CURRENTS

Drift Deviation in Currents

If you are navigating an easterly course and the current is flowing from the south, you will slowly begin to drift northward with the current, causing you to swim off course. The formula we will discuss is simplified so it can be understood and applied by the average diver. It will explain the concept of current deviation and allow you to navigate in a more accurate manner. It is possible to make a more accurate correction for the current, but it is an advanced navigational technique and is not covered in this manual. The factors for correcting current deviation are covered below.

Factors for Drift Correction Formula

To estimate the correction for current deviation, you first need to know a few basic factors:

- The approximate speed of the current.
- The distance to your destination.
- Your swimming speed.

■ **Current Speed:** The approximate speed of the current can be obtained from the dive leader or boat captain, or can be determined from tidal current tables and charts. The speed of the current can be estimated by timing a floating object over a set distance. For example, an object floating in a 1-knot current will move 1.69 feet (.52 metres) per second (see formula on page 61). This means it will travel 10 feet (3 metres) in about 6 seconds. If you are on a boat and you know the length of the boat, you can estimate the speed by timing an object as it moves past the boat. A 50-foot (15 metre) boat will take 30 seconds for an object to drift by in a 1-knot current.

■ **Distance to Destination:** The distance to your destination can be determined from charts, but for distances under one quarter mile (400 metres) an estimate should suffice. Remember you are looking to estimate, not calculate.

■ **Swimming Speed:** Your swimming speed can be determined before the dive by swimming a predetermined course of 100 feet (30 metres). Swim the 100 feet (30 metres) a few times to figure out the average number of seconds it takes to swim. For our example let's say it takes 33 seconds; this means your *feet-per-second rate* is 3. To figure out your rate, use the formula below:

Distance ÷ Seconds = Feet-Per-Second Rate

Another way to figure your swimming speed is by *feet-per-kick*. Swim the same 100 feet (30 metres), but this time, count your fin kicks instead of your time. If it takes you 25 kicks to cover 100 feet, this means your feet-per-kick rate is 4. To figure your rate, use the formula below:

> ## Distance ÷ Kicks = Feet-Per-Kick Rate

Once you have these three factors established (speed of current, distance to destination, swimming speed), you can figure the drift correction.

Formula for Drift Correction in Currents

For our example, let's say that we have already determined the following factors:

Example:

- *Speed of the current = 1 knot*
- *Distance to swim = 300 feet (91.5 metres)*
- *Swimming speed = 3 feet/second (.91 metres/second)*

To determine your corrected course, you will need to draw a simple diagram that will show your true course, deviated course, and corrected course.

For our example, let us say that we are swimming from point A, which is the boat, to point B, which is a wreck (Figure 4-8a, see page 60). The first step is to determine how far off course you will deviate if you do not correct for the current.

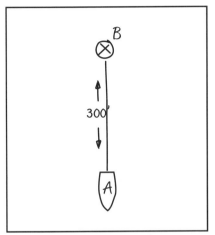

Figure 4-8a *Draw a simple diagram to show your course from point A to point B.*

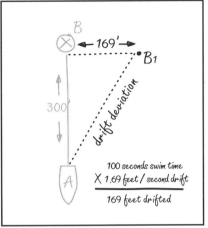

Figure 4-8b *Point B-1 designates how far you will drift according to the distance and speed of the current.*

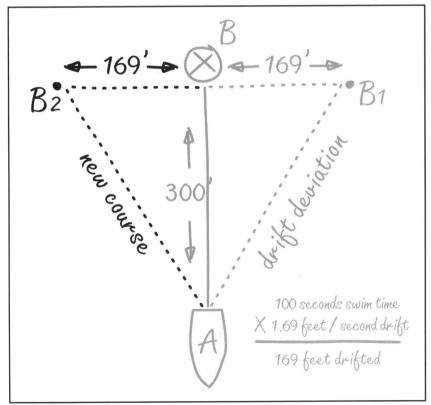

Figure 4-8c *Point B-2 designates your new course adjusting for current deviation.*

To begin with, we will figure out how long it will take you to swim to point B1. Remember that these are only approximations, because the distance from A to B is less than the distance from A to B1.

(Example, continued)

- **Swim speed = 3 ft. (.91 metre)/second**
- **Distance to swim = 300 feet (91.5 metres)**
- **Then 300 feet ÷ 3 feet/second = 100 seconds swim time**
  OR   91.5 metres ÷ .91 metres/second = 100 seconds

Now, because your total swimming time is measured in seconds, it will be easier to calculate your deviation if the current speed is also calculated into seconds (from nautical miles per hour):

(Example, continued)

- **1 nautical mile = 6076 feet (1852 metres)**
- **1 hour = 3600 seconds**
- **6076 feet ÷ 3600 seconds = 1.69 feet/second**
  OR   1852 metres ÷ 3600 seconds = .5 metres/second

This means that you will drift 1.69 feet (.5 metres) for every second you swim, so if you swim 100 seconds you will drift 169 feet (50 metres) total to point B1 (Figure 4-8b).

(Example, continued)

- **1.69 ft./second x 100 seconds**
  = 169 feet that you have deviated off course.

The final step is to figure out your new course, or point A to point B2 (Figure 4-8c). Now that you know that point B1 is 169 feet from point B, you know that point B2 is also 169 feet from point B in the other direction. Now you simply need to estimate the location of point B2 and take a compass heading for it. So for our example, if we know the wreck is due north from our boat, point B2 lies at 280 degrees on our compass.

As mentioned earlier, there are additional calculations you can make so that your time, distance and compass heading are more accurate, however, they are also more technical. These 3 simple steps shown here will give you an idea of how current affect navigation, and how to approximate your deviation so it can be corrected. Once again, you should remember that these steps need practical application. They are estimates that can be made in your head with relative ease once you know your swim speed, and that you will drift 6 feet (1.8 metres) every 10 seconds in a 1 knot current. Armed with this information you can make easy estimates that will help put you near your exit point rather than coming up in a place you really do not want to be.

Knowing how to find your way to and from your exit point is always important, but the need for navigation skills is increased in special situations such as when diving at night or in longshore currents.

However, navigation is not just a safety skill, it can also be a fun and challenging activity. In Chapter 5 we will look at how to participate in "the sport of navigation."

CHAPTER 4
REVIEW

1. When you are diving in turbid water, your instruments may provide your only sense of _____ and _____.

2. A compass is particularly handy at night because you must be more accurate about reaching your _____ point.

3. When deep diving, the last thing a diver should do is to get lost, or _____ for directions.

4. When navigating from a beach, a water condition that may cause some problems is a _____ current, or a current that runs near the shore.

5. Any time the waves reach the shore over a shallow reef or sand bar, that water usually returns to the sea via a _____ _____.

THE SPORT OF NAVIGATION
5

CHAPTER 5:
THE SPORT
OF NAVIGATION

Navigation is a skill that will be used and perfected throughout your diving career. The more you dive, the more you can practice advanced navigation techniques. In fact, navigation can even be considered somewhat of a sport. It is a challenging activity that allows you to continually test your own abilities, and to compete against others. You and your dive buddies, or your local dive club, can create friendly competitions to increase your skill and interest. In limited visibility, navigation can even add a whole new level of enjoyment and challenge to diving.

In this chapter we will look at some of the advanced activities that are part of the "sport" of navigation.

MULTIPLE COURSE HEADINGS

Once you have perfected a straight out-and-back reciprocal compass course, the next step is to move on to multiple course headings. In a multiple course, you can plan to make two or more stops on your compass course, and still make it back to your exit point. This gives you more options for diving, especially in limited visibility, where a compass is the best possible form of navigation.

3-Sided Course

A triangular course involves 3 compass headings, just as a triangle has 3 angles. The easiest course for a diver to calculate is a triangle with 3 equal turns of 120 degrees (Figure 5-1). However, any 3 angles can be used as long as they total 360 degrees.

As we mentioned in Chapter 2, it is always easiest to figure out your compass course ahead of time, while you are still on land. This may not always be feasible if you are not quite sure what you are going to do on your dive, such as when you are pleasure diving. In this case, you may want to take a slate with you under water so you can figure your compass calculations as you set them.

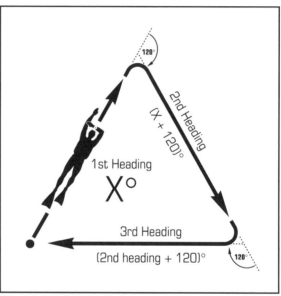

Figure 5-1 *A 3-sided or triangular compass course.*

4-Sided Course

If you would like to go on and expand your compass skills, the next step would be to try a four-point course. Another time would be if you are navigating around an object such as a large rock mound that is too tall to swim over.

Navigating a square course is easy because you simply make right angle, or 90 degree, turns. Four turns will equal 360 degrees and get you back to where you started (Figure 5-2).

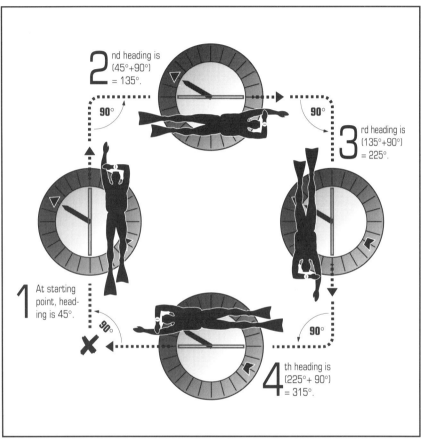

Figure 5-2   *Navigating a square course is easy because you simply make right angle, or 90 degree turns.*

Should you need to navigate around an object under water, it will be simple to do with a square pattern. You do not need to make any calculations on a slate to set a compass by 90 degrees, or at a right angle, especially when most compasses have the four directions of north, south, east, and west already marked at a 90 degree angle.

Let's say you begin your course on a 30 degree heading when you come upon an obstacle. Your first turn will need to be a 90 degree, left hand turn. By subtracting 90 degrees from 30, you will find that your course will be 300 degrees. When you come to the end of the obstruction you will need to turn again, this time a 90 degree turn to your right. By adding 90 degrees to 300 we see that our new course is again 30 degrees. To calculate your next right hand turn you must add another 90 degrees to your course of 30 degrees, for a course of 120 degrees. Once you have cleared the obstruction you are ready for your fourth 90 degree turn. This turn will be to your left. Again, subtract 90 degrees from your course of 120 degrees and you will be back to your original course of 30 degrees (Figure 5-3).

In order to reach your original course after clearing the obstruction, you will need to swim the same approximate fin kicks on each turn. For example, if the time you swim between your first and second turn is 25 seconds, then you will again need to swim that time between your third and

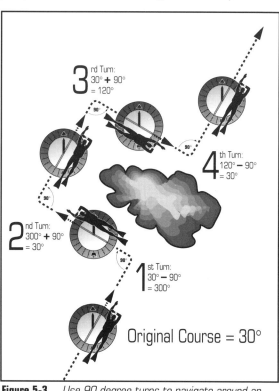

Figure 5-3 *Use 90 degree turns to navigate around an obstacle.*

final turn to be back on course. Time is used as a measurement since it is difficult or impossible to judge distance under water, and counting kicks can also be a problem.

If you were navigating a square course instead of swimming around an object, your final turn would have been to the right, instead of left, so you could end up at the same point where you started.

In order to make an exact square, you will always need to keep track of how long it takes you to swim the first leg of the square, and then swim each leg after that at the same rate. Otherwise you will end up swimming in a rectangle, or other 4 sided object, and not end up in the same place.

Create Fun and Competition

As we discussed earlier, you can make compass work fun by creating some friendly competitions. Use the multiple heading courses already discussed, or create your own. The fun you can have with navigation is only limited by your imagination, and the number of friends you can round up to participate.

One idea is to create a game by setting buoys that each have a clue attached. The diver is only given the compass heading to the first buoy, and must reach it in order to get the compass heading for the next buoy, and so on (Figure 5-4). The more limited the visibility is, the harder the game will be. The person who navigates to each buoy successfully wins the game.

You may want to make a variation of this game by putting a prize or other item you collect at each buoy. This not only proves the diver has reached it, but gives added incentive. One example is, for each marker that is picked

Figure 5-4   *One idea is to create a game by setting buoys that each have a clue attached.*

up at a buoy, the diver gets a playing card. The diver with the best poker hand then wins the game. The more markers you find, the more cards you can draw for your poker hand.

The games and competitions you can create are endless. The sport of navigation gives friends and dive clubs a fun way to enjoy the limited visibility season, or even a way to help raise money for your dive club or a charity.

DEAD RECKONING

Learning to navigate reciprocal courses, triangles and squares is great practice, but on real dives you will most likely want to change directions to follow the most interesting part of the reef, or the subject you wish to photograph. You will need to learn how to find your way back to the boat or shore after making many changes in your compass course.

Dead reckoning is defined as the calculation of one's position based on the distance run on various compass headings from the last precisely observed position, as accurate allowance as possible being made for currents, changes in speed, compass errors, etc.

For example, let us say that a diver swims approximately 50 yards (46 metres) into a slight current (Figure 5-5). The diver turns northeast for approximately 150 yards (137 metres) to swim after a ray. The diver then turns south and swims for another 100 yards to go take a picture of an old anchor. The diver then turns west and swims for approximately 75 yards (69 metres) to look for the boat.

At this point, the diver knows he is close to the boat, but also knows the slight current will have had some effect. He turns and swims approximately 30 yards (27 metres) north into the current to be sure of ascending in front of the boat.

Practice calculating where you are while driving in your car, or by using the compass on the boat while you are on the way to the dive site. This will help when having to position the boat for anchoring, or to help locate a reef or wreck.

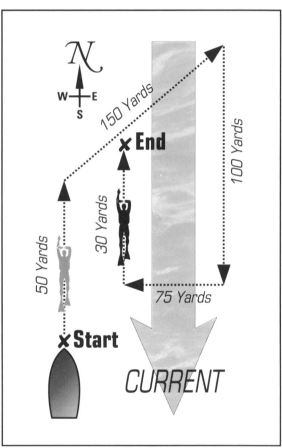

Figure 5-5 *Use dead reckoning to estimate your direction when a strict compass course is not followed.*

COMPASS EXPLORATION

Compass exploration is an even more advanced way to put your navigation skills to work. You need to be an advanced navigator and have some type of training in search and recovery before attempting an exploration on your own. However, even with training, you should only attempt to recover small items that can either be put in your pocket, or brought to the surface with a small lift bag (Figure 5-6). Recovering large items such as anchors and other parts of boats should be left to experienced salvage divers or professionals. To learn more about search and recovery, contact your local SSI store about a Specialty Course.

Figure 5-6   *You should only attempt to recover small items that can either be put in your pocket, or brought to the surface with a small lift bag.*

Search and recovery skills can be fun and useful, especially if you would like to recover that piece of equipment that fell overboard. In some cases you may be simply scouring the bottom for hidden treasure such as antique bottles or artifacts, other times you may be searching for a specific item such as the camera your buddy dropped on your last dive. The sport of compass exploration allows divers to combine search patterns with compass skills to explore an area based on the available visibility.

Define the Exploration Area

To begin with, any compass exploration will be more successful with a few basic rules. The first step is to define the search area. The more you can narrow it down, the better and faster your chance of finding what you are looking for. This step will not be as critical if you are simply scouring the bottom of the lake for fun.

Select the Best Search Pattern for Exploration

A successful underwater search can be conducted with two divers, a compass and a safety diver on shore (Figure 5-7, page 74). You will need to adjust the search pattern so each sweep is no wider than the limit of your visibility. If you would like to learn more about search patterns, we recommend that you take a *Search and Recovery* Specialty Course from you local SSI Authorized Dealer.

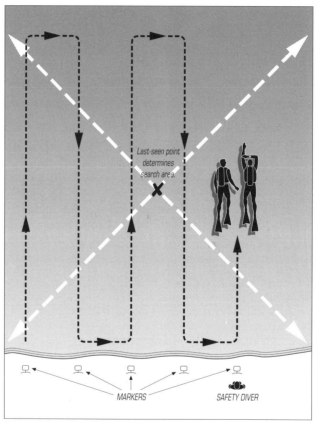

Figure 5-7   *Search technique using two divers, a compass, and a safety diver on shore.*

Establish a Tracking System

A tracking system will help you keep track of which areas have been searched, the type of pattern you used, and how closely the area was searched.

If you do not track your search area, you may end up duplicating your efforts by searching the same area twice. By keeping track of how closely the area was searched, you will know if you need to go back and re-dive areas that were broadly searched, or avoid areas that have already been looked over very closely.

Compass Exploration for Fun and Competition

The ability to run a search pattern can be valuable if you ever need to look for a piece of lost equipment, but this navigation technique can also provide some fun and competition for you and your friends.

Have fun by planting objects such as a barrel, or even a treasure chest full of goodies, and then having a contest to see which group can find it (or who can find it the fastest). Start with a simple course, and as the skill level of the group goes up, make the object more difficult to locate. There are all kinds of ways to enhance your abilities while having fun. Search and rescue patterns will not only improve your navigation skills, but they will also increase your overall diving ability. Remember, before attempting search and recovery patterns on your own, you should get some specialty training, or do a search with other divers who are already trained.

PLOTTING A BODY OF WATER

If you are diving locally, you may dive the same lakes or ocean waters month after month. One way to keep the same sites interesting is to plot them so you can explore the entire area. For example, if you al-ways dive along the shore or in the same bay, there may be a whole new world of adventure farther out, or in the middle of the lake.

Plotting is only realistic for small, shallow, isolated bodies of water, such as lakes, quarries, reservoirs or sinkholes (Figure 5-8). If you are an ocean diver, small bays or inlets could also be plotted. As we mentioned in the section on search and recovery, many

Figure 5-8 *Plotting is only realistic for small, shallow, isolated bodies of water.*

bodies of water have hidden wrecks, artifacts and other objects that would be of interest to a diver. You can even combine your exploration with specialized equipment such as underwater metal detectors.

Plotting the Lake with a Compass

By plotting a lake, you will actually be dividing it into workable sections that could be explored in one dive. How large your sections are will depend on the visibility, the terrain, and how closely you plan on exploring.

If you are plotting a lake that only has one good entry point, all of your compass headings should be taken from this point. If the body of water can be entered from any point along the shore, you actually plot it in equal vertical or horizontal sections. Either way, you should not plot a section that has no accessible entry point.

For our example, let's say you will be taking all of your headings from the dock because it is the only safe entry point. The first thing you will need to know is the visibility of the water, and how long it takes you to swim to the other side of the body of water. If it takes you 20 minutes and you are only planning on 45-minute dives, you should only count on swimming one sweep, up and back, on each dive. Now if the visibility is 10 feet (3 metres), you should take compass headings that are no more than 10 or 20 feet (3 or 6 metres) apart or you will not see the entire section on your dive.

The next step is to draw a rough map of the lake, or to use a real map if one is available. As you take your compass bearings, you will actually draw them onto the map for reference. With your map, you now have a complete record of your dives, so if you see any landmark or point of interest you would like to dive again, record it on your map. You can even chart depths and variances in the bottom. When you are done, you will have a "nautical chart" for your own local body of water. You can share the map with friends and your local dive club.

To actually take the bearings, simply stand at your entry point and face the direction you will be diving first. You can either start at the left or the right hand side of the lake. Point the lubber line at a reference point, or at the place you feel is about 10 feet (3 metres) from shore (remember, this was your distance between bearings because it was the limit of your visibility). Once the north arrow stops, set your index marks on the arrow. For this example, let's say the heading is 10 degrees.

Since you can assume that you will be able to swim approximately a new 10 degree course each dive, you can now break your map down into 10 degree quadrants (Figure 5-9a & b).

Now if your visibility turns out to be worse one day, you may need to split that quadrant, the section you have plotted, down into a 5 degree course. Conversely, if the visibility has improved, you may be able to see the length of more than one section, so you can increase your course to 15 or 20 degrees.

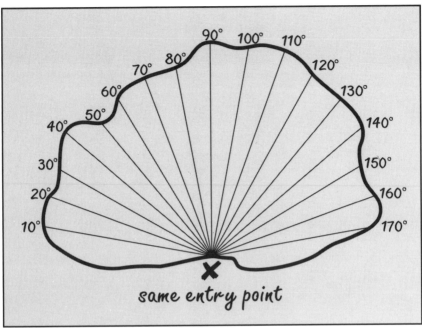

Figure 5-9a   *10° course from the same entry point.*

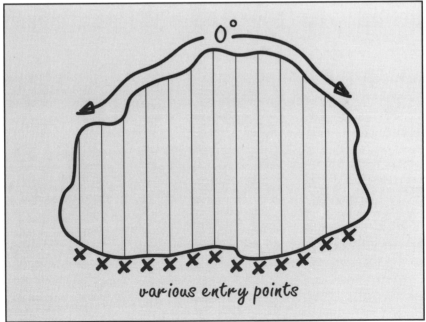

Figure 5-9b   *10- to 20-foot sections at 0° angle from different entry points.*

Your new navigation skills will open up new dive sites and possibilities you never would have had the ability to experience before!

SUMMARY

More than any other specialty, Navigation provides a wide variety of activities a diver can do in his or her own backyard. Local diving can be a challenging and rewarding experience. And even though the visibility may be limited, that doesn't mean your fun has to be. More than that, naviga-tion provides a sense of confidence and added safety no matter what body of water you are diving in.

When diving in the ocean, you have the added ability to negotiate currents, limited visibility, and other factors that could affect your dive. You have the flexibility to enjoy your dive, but always know that you can get back to the shore or boat without having to surface for directions. This makes all dives safer. Your new navigational skills will provide a whole world of potential, for they will open up new dive sites and possi-bilities you never would have had the ability to experience before!

CHAPTER 5
REVIEW

1. The easiest (3-sided) course for a diver to calculate is a triangle with 3 equal turns of 120 degrees. However, any 3 angles can be used as long as they total _____.

2. In order to make an exact square, you will always need to keep track of how long it takes you to swim the first leg of the square, and then swim each leg after that at _____ _____ _____.

3. The sport of compass exploration allows divers to combine _____ patterns with compass skills to explore an area based on the available _____.

4. The ability to run a search pattern can be valuable if you ever need to look for a piece of _____ _____, but this navigation technique can also provide some _____ and _____ for you and your friends.

5. One way to keep the same site interesting is to _____ it so you can explore the entire area.

APPENDIX

Equipment Checklist

Diver's Repair Kit

Log System Flow Chart

EQUIPMENT CHECKLIST

☐ Mask
☐ Snorkel & Keeper
☐ Fins
☐ Diving Suit
☐ Boots
☐ Gloves
☐ Hood
☐ Weight Belt
☐ Weights
☐ Buoyancy Compensator
☐ Backpack
☐ Tank(s) Full
☐ Regulator
☐ Alternate Air Source
☐ Pressure Gauge
☐ Watch or Timer
☐ Depth Gauge
☐ Compass
☐ Knife
☐ Whistle
☐ Decompression Computer
☐ Thermometer
☐ Defogging Solution
☐ Dive Light/Batteries
☐ Chemical Light

☐ Dive Flag
☐ Dive Tables
☐ Log Book
☐ Certification Card
☐ Speargun
☐ Extra Points
☐ Goody Bag
☐ Fishing License
☐ U/W Camera
☐ Flash or Strobe
☐ Batteries
☐ Film
☐ Slate
☐ Spare Parts Kit
☐ Swim Suit
☐ Towels
☐ Suntan Lotion/Sunscreen
☐ First Aid Kit
☐ Money for Emergency Calls
☐ Money for Air Fills
☐ Money for Galley & Tips
☐ Passport
☐ _____
☐ _____
☐ _____

SPARE PARTS/
REPAIR KIT

☐ Fin Straps & Buckles

☐ Mask Straps & Buckles

☐ Snorkel Keeper

☐ Knife Retaining Kit

☐ Knife Leg Strap

☐ Needle and Thread

☐ CO_2 Cartridges

☐ O-rings, Bulb for Light

☐ Batteries

☐ Dust Cap

☐ Regulator Port Plug

☐ Regulator Mouthpiece
and Cable Ties

☐ O-rings

☐ Silicone Spray

☐ Silicone Grease

☐ Wet Suit Cement

☐ BC Patch Kit

☐ Buckles for BC

☐ Buckle for Weight Belt

☐ Screwdriver
(Straight & Phillips)

☐ Pliers

☐ Crescent Wrench

☐ 5/32-Inch Allen Wrench

☐ WD-40®

☐ _____

☐ _____

☐ _____

☐ _____

☐ _____

☐ _____

ow SSI's Total DiveLog System Works...

FAME
RED
FIRE

Wealth

MARR

PINK
while
red

FAMILY
ELDERS
green

CHILD

METAL

KNOWLEDGE
SELF-CULTIVATION
blue

water
Black

Helpful
people

CAREER